Year 6 Mental Maths

30 tests, with answers, for home and school use

By Chris James

Version 1.0

Published on 1st September 2015

Contents

Introduction 1

How to Use This Book 1

    Completing the Tests Under Exam Conditions 1

    Completing the Tests Informally 2

Test 1 3

Test 1 Answers 10

Test 2 13

Test 2 Answers 20

Test 3 23

Test 3 Answers 30

Test 4 33

Test 4 Answers 40

Test 5 43

Test 5 Answers 50

Test 6 53

Test 6 Answers 60

Test 7 63

Test 7 Answers 70

Test 8 73

Test 8 Answers 80

Test 9 83

Test 9 Answers 90

Test 10 93

Test 10 Answers 100

Introduction

This book contains a series of 30 mental maths test for home and school use that are designed to improve your child's or class' mental maths skills. The tests are grouped into a series of three tests that each has similar questions. Question 1 of Test 1a is therefore very similar to question 1 in Test 1b and question 1 in Test 1c. This is to allow your child or children the chance to learn from their mistakes and try to improve on subsequent tests. I believe this opportunity to consolidate their knowledge is essential if you wish to see real progress. It is therefore really important that you take some time to explain to your child or class the way to correctly do each question at the end of each test, before they do the next test.

How to Use This Book

Each test is split into 3 parts: the test questions, the question sheets and the answer sheet. Each test has been written with the intention of it being completed under exam conditions, although it is possible for them to be used less formally.

Completing the Tests Under Exam Conditions

You will need to print off the correct number of question sheets. There are two question sheets to a page, so they will need to be cut up accordingly.

Before administering the test you will need to explain to your child or class that the questions get progressively harder. They will therefore have 5 seconds to answer questions 1 to 5, 10 seconds to answer questions 6 to 15 and 15 seconds to answer questions 16 to 20. There is some information written next to some of the answer boxes to help them with the answer. They should be encouraged to work out the answers in their heads, but they can write down further information if they so require.

You should then read out each question twice and your child or children should then attempt to answer the question. You will need a stopwatch to time how long you should leave between the end of one question and the beginning of the next. If you find that your child or class are struggling to answer the questions in the time given, you should adjust the time accordingly to give them more time to answer. The aim though should be for them to answer the questions within the time frame given in the previous paragraph. You could though give slightly longer time for tests a and b, and attempt test c in the recommended time.

Once the test has been completed you should take in the papers to mark them. Once they have been marked, it is very important, as mentioned previously, that you then take time to go through any mistakes that have been made by your child or class.

Completing the Tests Informally

For more formal use, you will need to provide everyone doing the test with a copy of the question paper and a copy of the answer paper. They can then read and answer the questions at their own pace. This removes the pressure of completing the work under time constraints.

As previously, it is vital that you mark the tests and give your child or class to correct any errors with your guidance.

Test 1

Test 1a

1 How many tens in seventy?

2 What is the total of 12 and 6?

3 What number is double thirty?

4 Subtract 12 from 20

5 What time is it half an hour before 9 o'clock?

6 What number is half of 74?

7 How many millimetres are there in one and a half litres?

8 Subtract 29p from £1

9 What number is half way between 32 and 40?

10 Subtract 3 ½ from 8?

11 How many sides does a hexagon have?

12 A chocolate bar costs 47p. What is the cost of 2 bars?

13 How many days are there in 5 weeks?

14 A Science lesson starts at 1:20 and finishes at 2:00. How long is the Science lesson?

15 What is £19.50 and £11.50?

16 Find the total of 15.5, 16.5 and 17

17 What is half of the product of 9 and 12?

18 Half of a number is 6.5

What is the number?

19 Divide 56 by 4?

20 A rectangle measures 5cm by 7cm.
What is the area of the rectangle?

	Test 1a Answer Sheet				Test 1a Answer Sheet	
	Help	Answer			Help	Answer
1	70		1	70	
2			2		
3	30		3	30	
4	20		4	20	
5	9 o'clock		5	9 o'clock	
6	74		6	74	
7		ml	7		ml
8	29p	p	8	29p	p
9			9		
10	3 ½		10	3 ½	
11			11		
12	47p	p	12	47p	p
13			13		
14	1:20pm	mins	14	1:20pm	mins
15		£	15		£
16	15.5, 16.5 & 17		16	15.5, 16.5 & 17	
17			17		
18	6.5		18	6.5	
19			19		
20		cm²	20		cm²

Test 1b

1   How many tens in eighty?

2   What is the total of 13 and 9?

3   What number is double forty?

4   Subtract 11 from 30

5   What time is it half an hour before 11 o'clock?

6   What number is half of 52?

7   How many millimetres are there in four and a half litres?

8   Subtract 39p from £1

9   What number is half way between 32 and 54?

10 Subtract 2 ½ from 10?

11 How many sides does a pentagon have?

12 A chocolate bar costs 27p. What is the cost of 2 bars?

13 How many days are there in 8 weeks?

14 A Science lesson starts at 3:25 and finishes at 4:05. How long is the Science lesson?

15 What is £29.50 and £14.50?

16 Find the total of 15.5, 26.5 and 27

17 What is half of the product of 9 and 8?

18 Half of a number is 6.25

    What is the number?

19 Divide 64 by 4?

20 A rectangle measures 9cm by 7cm.
    What is the area of the rectangle?

	Test 1b Answer Sheet	
	Help	Answer
1	80	
2		
3	40	
4	30	
5	11 o'clock	
6	52	
7		ml
8	39p	
9		
10	2 ½	
11		
12	27p	p
13		
14	3:25pm	mins
15		£
16	15.5, 26.5 & 27	
17		
18	6.25	
19		
20		cm²

	Test 1b Answer Sheet	
	Help	Answer
1	80	
2		
3	40	
4	30	
5	11 o'clock	
6	52	
7		ml
8	39p	
9		
10	2 ½	
11		
12	27p	p
13		
14	3:25pm	mins
15		£
16	15.5, 26.5 & 27	
17		
18	6.25	
19		
20		cm²

Test 1c

1 How many tens in ninety?

2 What is the total of 15 and 9?

3 What number is double fifty?

4 Subtract 11 from 40

5 What time is it two and half an hours before 4 o'clock?

6 What number is half of 92?

7 How many millimetres are there in three and a half litres?

8 Subtract 31p from £1

9 What number is half way between 32 and 50?

10 Subtract 4 ½ from 10?

11 How many sides does a parallelogram have?

12 A chocolate bar costs 38p. What is the cost of 2 bars?

13 How many days are there in 4 weeks?

14 A Science lesson starts at 3:25pm and finishes at 4:15. How long is the Science lesson?

15 What is £19.50 and £14.50?

16 Find the total of 25.5, 26.5 and 27

17 What is half of the product of 9 and 6?

18 Half of a number is 7.25

   What is the number?

19 Divide 76 by 4?

20 A rectangle measures 3cm by 9cm.
   What is the area of the rectangle?

	Test 1c Answer Sheet	
	Help	Answer
1	90	
2		
3	50	
4	40	
5	4 o'clock	
6	92	
7		ml
8	31p	
9		
10	4 ½	
11		
12	38p	P
13		
14	3:25pm	mins
15		£
16	25.5, 26.5 & 27	
17		
18	7.25	
19		
20		cm²

	Test 1c Answer Sheet	
	Help	Answer
1	90	
2		
3	50	
4	40	
5	4 o'clock	
6	92	
7		ml
8	31p	
9		
10	4 ½	
11		
12	38p	p
13		
14	3:25pm	mins
15		£
16	25.5, 26.5 & 27	
17		
18	7.25	
19		
20		cm²

Test 1 Answers

	Test 1a Answers	
	Help	Answer
1	70	7
2		18
3	30	60
4	20	8
5	9 o'clock	8:30
6	74	37
7		1500 ml
8	29p	71p
9		36
10	3 ½	4 ½
11		6
12	47p	94p
13		35
14	1:20pm	40 mins
15		£31
16	15.5, 16.5 & 17	49
17		54
18	6.5	13
19		14
20		35 cm²

	Test 1b Answers	
	Help	Answer
1	80	8
2		22
3	40	80
4	30	19
5	11 o'clock	10:30
6	52	26
7		4500 ml
8	39p	61p
9		43
10	2 ½	7 ½
11		5
12	27p	54p
13		56
14	3:25pm	40 mins
15		£44
16	15.5, 26.5 & 27	69
17		36
18	6.25	12.5
19		16
20		63 cm²

	Test 1c Answers	
	Help	Answer
1	90	9
2		24
3	50	100
4	40	29
5	4 o'clock	1:30 pm
6	92	46
7		3500 ml
8	31p	69p
9		41
10	4 ½	5 ½
11		4
12	38p	76 p
13		28
14	3:25pm	50 mins
15		£34
16	25.5, 26.5 & 27	79
17		27
18	7.25	14.5
19		19
20		27 cm²

Test 2

Test 2a

1. How many fives make 60?

2. What is nine multiplied by seven?

3. I have £100. I spend £69. How much change do I get?

4. How many millimetres are there in 4 ½ cm?

5. Multiply 83 by 10

6. What is £5.52 rounded to the nearest pound?

7. What fraction of the shape is coloured?

8. What number is double 28?

9. How many £10 notes do you need to have £350?

10. What is the difference between 89 and 301?

11. What is double 0.59?

12. How many minutes are there in two and a half hours?

13. What is 3.3 add 5.9?

14. What is the next number in this sequence: -7, -5, -3?

15. What is 1 - 0.01?

16. A rectangle is 9cm long and 5cm wide. What is its perimeter?

17. How many seconds are there in 30minutes?

18. If I save £3 every month, how much will I have after three years?

19. What number is 2/3 of 42?

20. How much is 25% of 180?

Test 2a Answer Sheet			Test 2a Answer Sheet		
	Help	Answer		Help	Answer
1			1		
2			2		
3	£100	£	3	£100	£
4		Mm	4		mm
5			5		
6		£	6		£
7			7		
8	28		8	28	
9	£350		9	£350	
10	89		10	89	
11	0.59		11	0.59	
12		mins	12		mins
13	3.3		13	3.3	
14	-7, -5, -3		14	-7, -5, -3	
15	0.01		15	0.01	
16		Cm	16		cm
17		s	17		s
18	£3	£	18	£3	£
19			19		
20	25%		20	25%	

Test 2b

1   How many fives make 75?

2   What is nine multiplied by eight?

3   I have £100. I spend £49. How much change do I get?

4   How many millimetres are there in 3 ½ cm?

5   Multiply 79 by 10

6   What is £5.42 rounded to the nearest pound?

7   What fraction of the shape is not coloured?

8   What number is double 38?

9   How many £10 notes do you need to have £450?

10   What is the difference between 79 and 201?

11   What is double 0.57?

12   How many minutes are there in three and a half hours?

13   What is 3.3 add 5.8?

14   What is the next number in this sequence: -8, -5, -2?

15   What is 1 - 0.03?

16   A rectangle is 9cm long and 6cm wide. What is its perimeter?

17   How many seconds are there in 40minutes?

18   If I save £2 every month, how much will I have after three years?

19   What number is 2/3 of 51?

20   How much is 25% of 240?

	Test 2b Answer Sheet				Test 2b Answer Sheet	
	Help	Answer			Help	Answer
1			1			
2			2			
3	£100	£	3		£100	£
4		Mm	4			mm
5			5			
6		£	6			£
7			7			
8	38		8		38	
9	£450		9		£450	
10	79		10		79	
11	0.57		11		0.57	
12		mins	12			mins
13	3.3		13		3.3	
14	-8, -5, -2		14		-8, -5, -2	
15	0.03		15		0.03	
16		Cm	16			cm
17		s	17			s
18	£2	£	18		£2	£
19			19			
20	25%		20		25%	

Test 2c

1   How many fives make 95?

2   What is six multiplied by eight?

3   I have £100. I spend £47. How much change do I get?

4   How many millimetres are there in 7 ½ cm?

5   Multiply 61 by 10

6   What is £9.65 rounded to the nearest pound?

7   What fraction of the shape is coloured?

8   What number is double 47?

9   How many £10 notes do you need to have £550?

10   What is the difference between 59 and 301?

11   What is double 0.58?

12   How many minutes are there in four and a half hours?

13   What is 3.5 add 5.7?

14   What is the next number in this sequence: -9, -5, -1?

15   What is 1 - 0.04?

16   A rectangle is 8cm long and 9cm wide. What is its perimeter?

17   How many seconds are there in 50minutes?

18   If I save £4 every month, how much will I have after three years?

19   What number is 2/3 of 39?

20   How much is 25% of 420?

	Test 2c Answer Sheet		
	Help	Answer	
1			
2			
3	£100	£	
4			Mm
5			
6		£	
7			
8	47		
9	£550		
10	59		
11	0.58		
12			mins
13	3.5		
14	-9, -5, -1		
15	0.04		
16			Cm
17			s
18	£4	£	
19			
20	25%		

	Test 2c Answer Sheet		
	Help	Answer	
1			
2			
3	£100	£	
4			mm
5			
6		£	
7			
8	47		
9	£550		
10	59		
11	0.58		
12			mins
13	3.5		
14	-9, -5, -1		
15	0.04		
16			cm
17			s
18	£4	£	
19			
20	25%		

Test 2 Answers

	Test 2a Answers	
	Help	Answer
1		12
2		63
3	£100	£31
4		45 mm
5		830
6		£6
7		5/8
8	28	56
9	£350	35
10	89	212
11	0.59	1.18
12		150 mins
13	3.3	9.2
14	-7, -5, -3	-1
15	0.01	0.99
16		28cm
17		1800s
18	£3	£108
19		28
20		45

	Test 2b Answers	
	Help	Answer
1		15
2		72
3	£100	£51
4		35 mm
5		790
6		£5
7		3/8
8	38	76
9	£450	45
10	79	122
11	0.57	1.14
12		210 mins
13	3.3	9.1
14	-8, -5, -2	1
15	0.03	0.97
16		30cm
17		2400s
18	£2	£72
19		34
20		60

	Test 2c Answers	
	Help	Answer
1		19
2		48
3	£100	£53
4		75 mm
5		610
6		£10
7		7/12
8	47	94
9	£550	55
10	59	242
11	0.58	1.16
12		270 mins
13	3.5	9.2
14	-9, -5, -1	3
15	0.04	0.96
16		34cm
17		3000s
18	£4	£144
19		26
20		105

Test 3

Test 3a

1   Divide 42 by 6

2   Subtract 13 from 30

3   Add 28 to 29

4   How many minutes are there in three hours?

5   What is ¾ of 20?

6   Add together 13, 14 and 15

7   What is 65 add 75?

8   A chocolate bar costs 65p. How much change would you get from £1?

9   What is half of 6.8?

10   Each side of a square is 2cm long. What is the perimeter of the square?

11   What is 105 divided by 5?

12   What is the mean of 14, 12 and 4?

13   What temperature is 10°C higher than -7°C?

14   What is half of 144?

15   What is 1.2 multiplied by 4?

16   A cube is 2.1cm long. 100 cubes are placed side by side in a line. How long is the line of cubes in cm?

17   A square has an area of 25cm². How long is each side?

18   One of the angles in an isosceles is 40°. What could the other 2 angles be?

19   Write the number between 110 and 120 whose digits add up to 9?

20   Write down four of the factors of 12

	Help	Answer
	Test 3a Answer Sheet	
	Help	Answer
1	42	
2	13	
3	28	
4		mins
5		
6		
7		
8	65p	p
9	6.8	
10		Cm
11	105	
12		
13	-7°C	°C
14	144	
15	1.2	
16	2.1	Cm
17		Cm
18	40°	A = °
		B = °
19		
20	12	

	Help	Answer
	Test 3a Answer Sheet	
	Help	Answer
1	42	
2	13	
3	28	
4		mins
5		
6		
7		
8	65p	p
9	6.8	
10		cm
11	105	
12		
13	-7°C	°C
14	144	
15	1.2	
16	2.1	cm
17		cm
18	40°	A = °
		B = °
19		
20	12	

Test 3b

1 Divide 42 by 7

2 Subtract 16 from 30

3 Add 38 to 29

4 How many minutes are there in four hours?

5 What is ¾ of 12?

6 Add together 23, 24 and 25

7 What is 65 add 85?

8 A chocolate bar costs 75p. How much change would you get from £1?

9 What is half of 5.8?

10 Each side of a square is 4cm long. What is the perimeter of the square?

11 What is 125 divided by 5?

12 What is the mean of 16, 13 and 7?

13 What temperature is 10°C higher than -6°C?

14 What is half of 166?

15 What is 1.25 multiplied by 4?

16 A cube is 2.4cm long. 100 cubes are placed side by side in a line. How long is the line of cubes in cm?

17 A square has an area of 49cm². How long is each side?

18 One of the angles in an isosceles is 50°. What could the other 2 angles be?

19 Write the number between 120 and 130 whose digits add up to 8?

20 Write down four of the factors of 16

	Test 3b Answer Sheet				Test 3b Answer Sheet	
	Help	Answer			Help	Answer
1	42		1		42	
2	16		2		16	
3	38		3		38	
4		mins	4			mins
5			5			
6			6			
7			7			
8	75p	p	8		75p	p
9	5.8		9		5.8	
10		Cm	10			cm
11	125		11		125	
12			12			
13	-6°C	°C	13		-6°C	°C
14	166		14		166	
15	1.25		15		1.25	
16	2.4cm	Cm	16		2.4	cm
17		Cm	17			cm
18	50°	A = °	18		50°	A = °
		B = °				B = °
19			19			
20	16		20		16	

Test 3c

1   Divide 56 by 7

2   Subtract 17 from 30

3   Add 38 to 23

4   How many minutes are there in six hours?

5   What is ¾ of 16?

6   Add together 22, 23 and 24

7   What is 65 add 95?

8   A chocolate bar costs 73p. How much change would you get from £1?

9   What is half of 5.4?

10   Each side of a square is 7cm long. What is the perimeter of the square?

11   What is 115 divided by 5?

12   What is the mean of 15, 14 and 10?

13   What temperature is 15°C higher than -6°C?

14   What is half of 168?

15   What is 1.3 multiplied by 4?

16   A cube is 2.7cm long. 100 cubes are placed side by side in a line. How long is the line of cubes in cm?

17   A square has an area of 81cm². How long is each side?

18   One of the angles in an isosceles is 30°. What could the other 2 angles be?

19   Write the number between 120 and 130 whose digits add up to 5?

20   Write down four of the factors of 24

	Test 3c Answer Sheet				Test 3c Answer Sheet	
	Help	Answer			Help	Answer
1	56		1	56		
2	17		2	17		
3	38		3	38		
4		mins	4			mins
5			5			
6			6			
7			7			
8	73p	p	8	73p		p
9	5.4		9	5.4		
10		Cm	10			cm
11	115		11	115		
12			12			
13	-6 °C	°C	13	-6 °C		°C
14	168		14	168		
15	1.3		15	1.3		
16	2.7cm	Cm	16	2.7cm		cm
17		Cm	17			cm
18	30°	A = °	18	30°		A = °
		B = °				B = °
19			19			
20	24		20	24		

Test 3 Answers

	Test 3a Answers	
	Help	Answer
1	42	7
2	13	17
3	28	57
4		180 mins
5		15
6		42
7		140
8	65p	35p
9	6.8	3.4
10		8 cm
11	105	21
12		10
13	-7°C	3°C
14	144	72
15	1.2	4.8
16	2.1	210 cm
17		5cm
18	If A = 40°, then B= 100° If A = 70°, then B = 70° Or the other way round	
19		117
20	Any four out of 1, 2, 3, 4, 6 and 12	

	Test 3b Answers	
	Help	Answer
1	42	6
2	16	14
3	38	67
4		240 mins
5		9
6		72
7		150
8	75p	25p
9	5.8	2.9
10		16 cm
11	125	25
12		12
13	-6°C	4°C
14	166	83
15	1.25	5
16	2.4	240 cm
17		7cm
18	If A = 50°, then B= 80° If A = 65°, then B = 65° Or the other way round	
19		125
20	Any four out of 1, 2, 4, 8 and 16	

	Test 3c Answers	
	Help	**Answer**
1	56	8
2	17	13
3	38	61
4		360 mins
5		12
6		69
7		160
8	73p	27p
9	5.4	2.7
10		28 cm
11	115	23
12		13
13	$-6\,^{\circ}C$	$9\,^{\circ}C$
14	168	84
15	1.3	5.2
16	2.7cm	270 cm
17		9cm
18	If A = 30°, then B= 120° If A = 75°, then B = 75° Or the other way round	
19		122
20	Any four out of 1, 2,3, 4, 6, 8,12 and 24	

Test 4

Test 4a

1   Write the next term in this sequence: 6, 8, 10, 12

2   Write the number three thousand, four hundred and twelve in figures

3   Multiply 45 by 10

4   What is 56 divided by 7?

5   What is the total of 18 and 23?

6   What is 42 multiplied by 5?

7   What is the remainder when you divide 47 by 5?

8   How many minutes are there in one and a quarter hours?

9   What is 15% of 1000?

10 If n = 10, what is 3n + 4?

11 I drink 350ml from a two-litre bottle. How much is left in millimetres?

12 How many years are there in seven decades?

13 What is double 18.5?

14 What must be added to 4^2 to make 50?

15 What is 125 + 225 + 325?

16 What is three-eighths of 48?

17 A camera at half price costs £135. What was the full price?

18 A train leaves at 8:45 a.m. and arrives at 10:35 a.m. How long does the journey take?

19 A t-shirt costs £28. It is reduced by 25%. What is the new price?

20 Two cereal bars cost 65p altogether. How much will it cost to buy four cereal bars?

	Test 4a Answer Sheet				Test 4a Answer Sheet	
	Help	Answer			Help	Answer
1			1			
2			2			
3	45		3		45	
4	56		4		56	
5			5			
6	42		6		42	
7	47		7		47	
8		mins	8			mins
9	15%		9		15%	
10	3n + 4		10		3n + 4	
11	350ml	ml	11		350ml	ml
12			12			
13	18.5		13		18.5	
14			14			
15			15			
16	3/8		16		3/8	
17	£135	£	17		£135	£
18	8:45a.m.		18		8:45a.m.	
19	£28	£	19		£28	£
20	65p	£	20		65p	£

Test 4b

1   Write the next term in this sequence: 5, 8, 11, 14

2   Write the number four thousand, two hundred and thirteen in figures

3   Multiply 49 by 10

4   What is 63 divided by 7?

5   What is the total of 19 and 23?

6   What is 43 multiplied by 5?

7   What is the remainder when you divide 39 by 5?

8   How many minutes are there in two and a quarter hours?

9   What is 15% of 500?

10   If n = 5, what is 3n + 7?

11   I drink 450ml from a two-litre bottle. How much is left in millimetres?

12   How many years are there in six decades?

13   What is double 19.5?

14   What must be added to 6^2 to make 50?

15   What is 25 + 125 + 325?

16   What is three-eighths of 56?

17   A camera at half price costs £145. What was the full price?

18   A train leaves at 8:55 a.m. and arrives at 10:35 a.m. How long does the journey take?

19   A t-shirt costs £32. It is reduced by 25%. What is the new price?

20   Two cereal bars cost 75p altogether. How much will it cost to buy four cereal bars?

	Test 4b Answer Sheet				Test 4b Answer Sheet	
	Help	Answer			Help	Answer
1			1			
2			2			
3	49		3		49	
4	63		4		63	
5			5			
6	43		6		43	
7	39		7		39	
8		mins	8			mins
9	15%		9		15%	
10	3n + 7		10		3n + 7	
11	450ml	ml	11		450ml	ml
12			12			
13	19.5		13		19.5	
14			14			
15			15			
16	3/8		16		3/8	
17	£145	£	17		£145	£
18	8:55a.m.		18		8:55a.m.	
19	£32	£	19		£32	£
20	75p	£	20		75p	£

Test 4c

1   Write the next term in this sequence: 5, 9, 13, 17

2   Write the number four thousand and thirteen in figures

3   Multiply 59 by 10

4   What is 84 divided by 7?

5   What is the total of 19 and 26?

6   What is 33 multiplied by 5?

7   What is the remainder when you divide 28 by 5?

8   How many minutes are there in one and three quarter hours?

9   What is 15% of 2000?

10 If $n = 6$, what is $3n + 3$?

11 I drink 550ml from a two-litre bottle. How much is left in millimetres?

12 How many years are there in nine decades?

13 What is double 22.5?

14 What must be added to 7^2 to make 100?

15 What is 425 + 125 + 325?

16 What is three-eighths of 96?

17 A camera at half price costs £155. What was the full price?

18 A train leaves at 8:55 a.m. and arrives at 10:25 a.m. How long does the journey take?

19 A t-shirt costs £24. It is reduced by 25%. What is the new price?

20 Two cereal bars cost 55p altogether. How much will it cost to buy four cereal bars?

	Test 4c Answer Sheet				Test 4c Answer Sheet	
	Help	Answer			Help	Answer
1				1		
2				2		
3	59			3	59	
4	84			4	84	
5				5		
6	33			6	33	
7	28			7	28	
8		mins		8		mins
9	15%			9	15%	
10	3n + 3			10	3n + 3	
11	550ml	ml		11	550ml	ml
12				12		
13	22.5			13	22.5	
14				14		
15				15		
16	3/8			16	3/8	
17	£155	£		17	£155	£
18	8:55a.m.			18	8:55a.m.	
19	£24	£		19	£24	£
20	55p	£		20	55p	£

Test 4 Answers

Test 4a Answers		
	Help	Answer
1		14
2		3412
3	45	450
4	56	8
5		41
6	42	210
7	47	2
8		75 mins
9	15%	150
10	3n + 4	34
11	350ml	1650 ml
12		70
13	18.5	37
14		34
15		675
16	3/8	18
17	£135	£270
18	8:45a.m.	1hr and 50mins
		or 110mins
19	£28	£21
20	65p	£1.30

	Help	Answer
	Test 4b Answers	
1		17
2		4213
3	49	490
4	63	9
5		42
6	43	215
7	39	4
8		135 mins
9	15%	75
10	3n + 7	22
11	450ml	1550 ml
12		60
13	19.5	39
14		14
15		475
16	3/8	21
17	£145	£290
18	8:55a.m.	1hr and 40mins
		or 100mins
19	£32	£24
20	75p	£1.50

	Test 4c Answers	
	Help	Answer
1		21
2		4013
3	59	590
4	84	12
5		45
6	33	165
7	28	3
8		105 mins
9	15%	300
10	3n + 3	21
11	550ml	1450 ml
12		90
13	22.5	45
14		51
15		875
16	3/8	36
17	£155	£310
18	8:55a.m.	1hr and 30mins
		or 90mins
19	£24	£18
20	55p	£1.10

Test 5

Test 5a

1   What is 13 add 18?

2   What is 5^2?

3   Subtract 35 from 120

4   How much is 7 times 8?

5   What is 36 divided by 4?

6   Write seven centimetres in metres.

7   A rectangle is 14cm long and 7cm wide. What is the area of the rectangle?

8   Pens cost £1.25 each. How much do four pens cost?

9   Multiply 3 by 7 and then add 8

10   Add 27 to 79

11   What percentage of the shape is coloured?

12   Subtract 1 ½ from 9

13   What is 76 more than 58?

14   What is double 361?

15   Add together 20, 30, 40, 50 and 60

16   A small pizza costs £3.50 and a large pizza costs £6.75
     What is the cost of 2 small pizzas and one large pizza?

17   A bucket holds 2 litres. A jug holds ¼ litre. How many jugs of water will
     fill the bucket?

18   I buy one item for £4.50 and one item for £7.50. How much change do I
     get from £20.00?

19   What number is exactly halfway between 10 and -20 ?

20   Two of the following numbers are prime numbers. Which are they?
     10, 11, 15, 16 and 19

	Test 5a Answer Sheet		
	Help		Answer
1	13		
2			
3	35		
4			
5			
6			m
7			cm²
8	£1.25		£
9	8		
10	27		
11			%
12	1 ½		
13	76		
14			
15	20,30,40,50 & 60		
16	Small: £3.50		
	Large: £6.75		
17	2 litres		
18			
19			
20	10, 11, 15, 16 and 19		

	Test 5a Answer Sheet		
	Help		Answer
1	13		
2			
3	35		
4			
5			
6			m
7			cm²
8	£1.25		£
9	8		
10	27		
11			%
12	1 ½		
13	76		
14			
15	20,30,40,50 & 60		
16	Small: £3.50		
	Large: £6.75		
17	2 litres		
18			
19			
20	10, 11, 15, 16 and 19		

Test 5b

1   What is 15 add 18?

2   What is 9^2?

3   Subtract 45 from 120

4   How much is 9 times 8?

5   What is 48 divided by 4?

6   Write two centimetres in metres.

7   A rectangle is 15cm long and 6cm wide. What is the area of the rectangle?

8   Pens cost £1.25 each. How much do eight pens cost?

9   Multiply 3 by 4 and then add 7

10   Add 25 to 68

11   What percentage of the shape is not coloured?

12   Subtract 1 ½ from 10

13   What is 77 more than 49?

14   What is double 472?

15   Add together 10, 30, 50, 70 and 90

16   A small pizza costs £2.50 and a large pizza costs £6.95
     What is the cost of 2 small pizzas and one large pizza?

17   A bucket holds 3 litres. A jug holds ¼ litre. How many jugs of water will fill the bucket?

18   I buy one item for £5.00 and one item for £6.50. How much change do I get from £20.00?

19   What number is exactly halfway between 10 and -4 ?

20   Two of the following numbers are prime numbers. Which are they?
     20, 21, 23, 26 and 29

		Test 5b Answer Sheet		
		Help	Answer	
	1	15		
	2			
	3	45		
	4			
	5			
	6			m
	7			cm²
	8	£1.25	£	
	9	7		
	10	25		
	11			%
	12	1 ½		
	13	77		
	14			
	15	10,30,50,70 & 90		
	16	Small: £2.50		
		Large: £6.95		
	17	3 litres		
	18			
	19			
	20	20, 21, 23, 26 and 29		

		Test 5b Answer Sheet		
		Help	Answer	
	1	15		
	2			
	3	45		
	4			
	5			
	6			m
	7			cm²
	8	£1.25	£	
	9	7		
	10	25		
	11			%
	12	1 ½		
	13	77		
	14			
	15	10,30,50,70 & 90		
	16	Small: £2.50		
		Large: £6.95		
	17	3 litres		
	18			
	19			
	20	20, 21, 23, 26 and 29		

Test 5c

1   What is 17 add 19?

2   What is 8^2?

3   Subtract 55 from 120

4   How much is 9 times 6?

5   What is 48 divided by 6?

6   Write eight centimetres in metres.

7   A rectangle is 13cm long and 5cm wide. What is the area of the rectangle?

8   Pens cost £1.25 each. How much do six pens cost?

9   Multiply 5 by 4 and then add 6

10   Add 34 to 69

11   What percentage of the shape is coloured?

12   Subtract 1 ½ from 8

13   What is 75 more than 48?

14   What is double 493?

15   A small pizza costs £3.25 and a large pizza costs £6.50
      What is the cost of 2 small pizzas and one large pizza?

16   Add together 10, 20, 50, 60 and 80

17   A bucket holds 4 litres. A jug holds ¼ litre. How many jugs of water will
      fill the bucket?

18   I buy one item for £8.25 and one item for £5.25. How much change do I
      get from £20.00?

19   What number is exactly halfway between 10 and -8 ?

20   Two of the following numbers are prime numbers. Which are they?
      30, 31, 33, 35 and 37

	Test 5c Answer Sheet		
	Help	Answer	
1	17		
2			
3	55		
4			
5			
6		m	
7		cm²	
8	£1.25	£	
9	6		
10	34		
11		%	
12	1 ½		
13	75		
14			
15	Small: £3.25		
	Large: £6.50		
16	10, 20, 50, 60 & 80		
17	4 litres		
18			
19			
20	30, 31, 33, 35 and 37		

	Test 5c Answer Sheet		
	Help	Answer	
1	17		
2			
3	55		
4			
5			
6		m	
7		cm²	
8	£1.25	£	
9	6		
10	34		
11		%	
12	1 ½		
13	75		
14			
15	Small: £3.25		
	Large: £6.50		
16	10, 20, 50, 60 & 80		
17	4 litres		
18			
19			
20	30, 31, 33, 35 and 37		

Test 5 Answers

	Test 5a Answers	
	Help	**Answer**
1	13	31
2		25
3	35	85
4		56
5		9
6		0.07m
7		98 cm²
8	£1.25	£ 5
9	8	29
10	27	106
11		75%
12	1 ½	7 ½
13	76	134
14		722
15	20,30,40,50 & 60	200
16	Small: £3.50	£13.75
	Large: £6.75	
17	2 litres	8
18		£8
19		-5
20	10, 11, 15, 16 and 19	11 and 19

	Test 5b Answers	
	Help	Answer
1	15	33
2		81
3	45	75
4		72
5		12
6		0.02m
7		90 cm²
8	£1.25	£ 10
9	7	19
10	25	93
11		25%
12	1 ½	8 ½
13	77	126
14		944
15	10,30,50,70 & 90	250
16	Small: £2.50	£11.95
	Large: £6.95	
17	3 litres	12
18		£8.50
19		3
20	20, 21, 23, 26 and 29	23 and 29

	Test 5c Answers	
	Help	**Answer**
1	17	36
2		64
3	55	65
4		54
5		8
6		0.08m
7		65 cm^2
8	£1.25	£ 7.50
9	6	26
10	34	103
11		30%
12	1 ½	6 ½
13	75	123
14		986
15	Small: £3.25	£13.00
	Large: £6.50	
16	10, 20, 50, 60 & 80	220
17	4 litres	16
18		£6.50
19		1
20	30, 31, 33, 35 and 37	31 and 37

Test 6

Test 6a

1   What is the sum of 3, 5 and 7?

2   Barry eats 1/3 of a pizza. How much pizza remains?

3   What is half of 72?

4   What are 9 lots of 9?

5   Subtract 99 from 345

6   Add 20p to £6.95

7   What is 30% of 50p?

8   What number is 3/4 of 60?

9   How many halves make 3 1/2?

10   What is 316 subtract 105?

11   Write the fraction four-tenths in its simplest form

12   Write the prime number closest to 10

13   When g = 7, what's the value of g x g + 1

14   Write three thousand and sixty seven grams in kilograms

15   A rectangle is 9cm long and 3 cm wide. What is the perimeter?

16   What is 24 multiplied by 8?

17   What number is exactly halfway between 1.2 and 2.2?

18   Add together 13, 23 and 33

19   I spill 0.8litres from a one litre bottle of water.
     How many millimetres remain?

20   Two of the following numbers are factors of 24. Which are they?
     3, 5, 9, 12 and 16

	Test 6a Answer Sheet				Test 6a Answer Sheet	
	Help	Answer			Help	Answer
1			1			
2	1/3		2		1/3	
3	72		3		72	
4			4			
5	345		5		345	
6	£6.95		6		£6.95	
7	30%	p	7		30%	p
8	3/4		8		3/4	
9			9			
10	316		10		316	
11	4/10		11		4/10	
12			12			
13	g x g + 1		13		g x g + 1	
14		kg	14			kg
15		cm	15			cm
16	24		16		24	
17	1.2 & 2.2		17		1.2 & 2.2	
18	13, 23 and 33		18		13, 23 and 33	
19	0.8litres		19		0.8litres	ml
20	3, 5, 9, 12 and 16		20		3, 5, 9, 12 and 16	

Test 6b

1   What is the sum of 4, 5 and 6?

2   Barry eats 3/5 of a pizza. How much pizza remains?

3   What is half of 52?

4   What are 9 lots of 8?

5   Subtract 99 from 366

6   Add 20p to £5.85

7   What is 40% of 50p?

8   What number is 3/4 of 80?

9   How many halves make 4 1/2?

10   What is 323 subtract 105?

11   Write the fraction six-tenths in its simplest form

12   Write the prime number closest to 20

13   When g = 6, what's the value of g x g + 3

14   Write two thousand and fifty seven grams in kilograms

15   A rectangle is 8cm long and 5 cm wide. What is the perimeter?

16   What is 22 multiplied by 8?

17   What number is exactly halfway between 1.4 and 2.4?

18   Add together 14, 24 and 34

19   I spill 0.7litres from a one litre bottle of water.
      How many millimetres remain?

20   Two of the following numbers are factors of 32. Which are they?
      3, 4, 9, 12 and 16

	Test 6b Answer Sheet				Test 6b Answer Sheet	
	Help	Answer			Help	Answer
1			1			
2	3/5		2		3/5	
3	52		3		52	
4			4			
5	366		5		366	
6	£5.85		6		£5.85	
7	40%	p	7		40%	p
8	3/4		8		3/4	
9			9			
10	323		10		323	
11	6/10		11		6/10	
12			12			
13	g x g + 3		13		g x g + 3	
14		kg	14			kg
15		cm	15			cm
16	22		16		22	
17	1.4 & 2.4		17		1.4 & 2.4	
18	14, 24 and 34		18		14, 24 and 34	
19	0.7litres		19		0.7litres	ml
20	3, 4, 9, 12 and 16		20		3, 4, 9, 12 and 16	

Test 6c

1. What is the sum of 2, 4 and 8?

2. Barry eats 3/4 of a pizza. How much pizza remains?

3. What is half of 56?

4. What are 9 lots of 7?

5. Subtract 99 from 406

6. Add 20p to £9.85

7. What is 20% of 50p?

8. What number is 3/4 of 84?

9. How many halves make 5 1/2?

10. What is 411 subtract 105?

11. Write the fraction six-ninths in its simplest form

12. Write the prime number closest to 25

13. When g = 8, what's the value of g x g + 4

14. Write three thousand and fifty six grams in kilograms

15. A rectangle is 9cm long and 5 cm wide. What is the perimeter?

16. What is 25 multiplied by 8?

17. What number is exactly halfway between 2.45 and 3.45?

18. Add together 16, 26 and 36

19. I spill 0.6litres from a one litre bottle of water.
    How many millimetres remain?

20. Two of the following numbers are factors of 40. Which are they?
    3, 4, 8, 12 and 17

	Test 6c Answer Sheet	
	Help	Answer
1		
2	3/4	
3	56	
4		
5	406	
6	£9.85	
7	20%	p
8	3/4	
9		
10	411	
11	6/9	
12		
13	g x g + 4	
14		kg
15		cm
16	25	
17	2.45 & 3.45	
18	16, 26 and 36	
19	0.6litres	
20	3, 4, 8, 12 and 17	

	Test 6c Answer Sheet	
	Help	Answer
1		
2	3/4	
3	56	
4		
5	406	
6	£9.85	
7	20%	p
8	3/4	
9		
10	411	
11	6/9	
12		
13	g x g + 4	
14		kg
15		cm
16	25	
17	2.45 & 3.45	
18	16, 26 and 36	
19	0.6litres	ml
20	3, 4, 8, 12 and 17	

Test 6 Answers

	Test 6a Answers	
	Help	Answer
1		15
2	1/3	2/3
3	72	36
4		81
5	345	246
6	£6.95	£7.15
7	30%	15p
8	¾	45
9		7
10	316	211
11	4/10	2/5
12		11
13	g x g + 1	50
14		3.067kg
15		24cm
16		192
17	1.2 & 2.2	1.7
18	13, 23 and 33	69
19	0.8litres	200ml
20	3, 5, 9, 12 and 16	3 and 12

	Test 6b Answers	
	Help	Answer
1		15
2	3/5	2/5
3	52	26
4		72
5	366	267
6	£5.85	£6.05
7	40%	20p
8	¾	60
9		9
10	323	218
11	6/10	3/5
12		19
13	g x g + 3	39
14		2.057kg
15		26cm
16	22	176
17	1.4 & 2.4	1.9
18	14, 24 and 34	72
19	0.7litres	300ml
20	3, 4, 9, 12 and 16	4 and 16

	Help	Answer
	Test 6c Answers	
1		14
2	¾	1/4
3	56	28
4		63
5	406	307
6	£9.85	£10.05
7	20%	10p
8	¾	63
9		11
10	411	306
11	6/9	2/3
12		23
13	g x g + 4	68
14		3.056kg
15		28cm
16	25	200
17	2.45 & 3.45	2.95
18	16, 26 and 36	78
19	0.6litres	400ml
20	3, 4, 8, 12 and 17	4 and 8

Test 7

Test 7a

1　Write twenty three thousand and twenty five in figures

2　Multiply sixteen by ten

3　What is the next number in the sequence: nine, thirteen, seventeen, twenty one?

4　Decrease twenty five by sixteen.

5　How much is half of fifty?

6　Add together six hundred, five hundred and four hundred.

7　What number is three quarters of thirty six?

8　What is the total of three, thirteen and twenty three?

9　How many metres are there in half a kilometre?

10　What is 3.5 divided by 10?

11　A pencil costs 45p. What is the cost of three pencils?

12　Write a multiple of forty that is greater than two hundred and fifty

13　What is the remainder when 59 is divided by 7?

14　Find the value of double 6.8

15　How many lengths of 10cm can be cut from a piece of string that is 67cm long?

16　How much is thirteen squared?

17　Halve sixty six and then halve the result

18　A rectangle is 6cm long and 2.5cm wide. What is the area?

19　I start with twenty pounds. First I spend thirteen pounds thirty nine pence and then two pounds and six pence.
How much change do I get from twenty pounds?

20　Multiply forty four by eleven

Test 7a Answer Sheet		
	Help	Answer
1		
2	16	
3	9,13, 17, 21	
4	25	
5		
6		
7	3/4	
8		
9		m
10	3.5	
11	45p	
12	250	
13	59	
14		
15		
16		
17	66	
18		cm²
19	£20	
20	44	

Test 7a Answer Sheet		
	Help	Answer
1		
2	16	
3	9,13, 17, 21	
4	25	
5		
6		
7	3/4	
8		
9		m
10	3.5	
11	45p	
12	250	
13	59	
14		
15		
16		
17	66	
18		cm²
19	£20	
20	44	

Test 7b

1   Write twenty two thousand and ninety nine in figures

2   Multiply eighteen by ten

3   What is the next number in the sequence: eight, eleven, fourteen, seventeen?

4   Decrease twenty four by sixteen.

5   How much is half of seventy?

6   Add together seven hundred, six hundred and five hundred.

7   What number is three fifths of forty five?

8   What is the total of four, fourteen and twenty four?

9   How many metres are there in one and a half kilometres?

10   What is 3.9 divided by 10?

11   A pencil costs 55p. What is the cost of three pencils?

12   Write a multiple of forty that is greater than three hundred and fifty

13   What is the remainder when 59 is divided by 6?

14   Find the value of double 6.7

15   How many lengths of 10cm can be cut from a piece of string that is 73cm long?

16   How much is fifteen squared?

17   Halve seventy and then halve the result

18   A rectangle is 8cm long and 2.5cm wide. What is the area?

19   I start with twenty pounds. First I spend twelve pounds thirty nine pence and then I spend seventy pence.
How much change do I get?

20   Multiply thirty four by eleven

	Test 7b Answer Sheet		
	Help	Answer
1		
2	18	
3	8,11, 14, 17	
4	24	
5		
6		
7	3/5	
8		
9		m
10	3.9	
11	55p	
12	350	
13	59	
14		
15		
16		
17	70	
18		cm²
19	£20	
20	34	

	Test 7b Answer Sheet		
	Help	Answer
1		
2	18	
3	8,11, 14, 17	
4	24	
5		
6		
7	3/5	
8		
9		m
10	3.9	
11	55p	
12	350	
13	59	
14		
15		
16		
17	70	
18		cm²
19	£20	
20	34	

Test 7c

1   Write thirty one thousand and ninety two in figures

2   Multiply nineteen by ten

3   What is the next number in the sequence: eight, fourteen, twenty, twenty six?

4   Decrease twenty three by sixteen.

5   How much is half of ninety?

6   Add together eight hundred, seven hundred and six hundred.

7   What number is three sevenths of forty two?

8   What is the total of five, fifteen and twenty five?

9   How many metres are there in two and a half kilometres?

10   What is 4.9 divided by 10?

11   A pencil costs 65p. What is the cost of three pencils?

12   Write a multiple of forty that is greater than four hundred and twenty

13   What is the remainder when 29 is divided by 6?

14   Find the value of double 7.7

15   How many lengths of 10cm can be cut from a piece of string that is 73cm long?

16   How much is fourteen squared?

17   Halve eighty two and then halve the result

18   A rectangle is 8cm long and 3.5cm wide. What is the area?

19   I start with twenty pounds. First I spend thirteen pounds forty nine pence and then I spend one pound and fifteen pence.
How much change do I get?

20   Multiply fifty five by eleven

	Test 7c Answer Sheet		
	Help	Answer	
1			
2	19		
3	8, 14, 20, 26		
4	23		
5			
6			
7	3/7		
8			
9			m
10	4.9		
11	65p		
12	420		
13	29		
14			
15			
16			
17	82		
18			cm^2
19	£20		
20	55		

	Test 7c Answer Sheet		
	Help	Answer	
1			
2	19		
3	8, 14, 20, 26		
4	23		
5			
6			
7	3/7		
8			
9			m
10	4.9		
11	65p		
12	420		
13	29		
14			
15			
16			
17	82		
18			cm^2
19	£20		
20	55		

Test 7 Answers

	Test 7a Answers	
	Help	Answer
1		23025
2	16	160
3	9,13, 17, 21	25
4	25	9
5		25
6		1500
7	¾	27
8		39
9		500m
10	3.5	0.35
11	45p	£1.35 or 135p
12	250	280, 320 etc
13	59	3
14		13.6
15		6
16		169
17	66	16.5
18		15 cm²
19	£20	£4.55
20	44	484

	Test 7b Answers	
	Help	Answer
1		22099
2	18	180
3	8,11, 14, 17	20
4	24	8
5		35
6		1800
7	3/5	27
8		42
9		1500m
10	3.9	0.39
11	55p	£1.65 or 165p
12	350	360, 400 etc
13	59	5
14		13.4
15		7
16		225
17	70	17.5
18		20 cm²
19	£20	£6.91
20	34	374

	Test 7c Answers	
	Help	**Answer**
1		31092
2	19	190
3	8,14, 20, 26	32
4	23	7
5		45
6		2100
7	3/7	18
8		45
9		2500m
10	4.9	0.49
11	65p	£1.95 or 195p
12	420	440, 480 etc
13	29	5
14		15.4
15		7
16		196
17	52	20.5
18		28 cm^2
19	£20	£5.36
20	55	605

Test 8

Test 8a

1   How many threes are there in twenty four?

2   Find the total of 13 and 27

3   What is the value of 7 x 8?

4   Share one hundred between four

5   How many days are there in April?

6   What is the square root of 25?

7   Write down a prime number that is bigger than 15 but less than 20

8   How many sevens are there in three thousand five hundred?

9   What is the answer when seventy is added to three thousand and forty six?

10   What is 4.9 divided by 10?

11   What is 25% of £5?

12   What is 2 hours and 15 minutes after 7:55a.m?

13   What is 1.4 multiplied by 3?

14   A temperature of 6°C is recorded. The temperature drops by 15°C.
     What is the new temperature?

15   Twelve cakes cost three pounds and twenty pence.
     How much do six cakes cost?

16   One angle in a triangle is 45° and another is 65°
     What is the size of the third angle?

17   What number is exactly half way between twenty four and seventy four?

18   What is the value of 4y + 5, when y = 3 ?

19   Write **one** multiple of 4 that is bigger than 145 but smaller than 155

20   How much is double 18.6?

	Test 8a Answer Sheet		
	Help	Answer
1		
2	13	
3		
4	100	
5		
6		
7		
8	3500	
9	70	
10	4.9	
11	25%	
12	7:55a.m	
13	1.4	
14	6°C	
15	£3.20	
16	45°	
17	24 & 74	
18	4y + 5	
19		
20	18.6	

	Test 8a Answer Sheet		
	Help	Answer
1		
2	13	
3		
4	100	
5		
6		
7		
8	3500	
9	70	
10	4.9	
11	25%	
12	7:55a.m	
13	1.4	
14	6°C	
15	£3.20	
16	45°	
17	24 & 74	
18	4y + 5	
19		
20	18.6	

Test 8b

1   How many threes are there in twenty seven?

2   Find the total of 14 and 36

3   What is the value of 9 x 8?

4   Share one hundred between five

5   How many days are there in September?

6   What is the square root of eighty one?

7   Write down a prime number that is bigger than 10 but less than 15

8   How many sevens are there in four thousand two hundred?

9   What is the answer when eighty is added to three thousand and forty one?

10   What is 7.5 divided by 10?

11   What is 25% of £6?

12   What is 2 hours and 25 minutes after 7:50a.m?

13   What is 1.5 multiplied by 3?

14   A temperature of 7°C is recorded. The temperature drops by 15°C.
     What is the new temperature?

15   Twelve cakes cost three pounds and thirty pence.
     How much do six cakes cost?

16   One angle in a triangle is 45° and another is 85°
     What is the size of the third angle?

17   What number is exactly half way between twenty two and seventy two?

18   What is the value of 3y + 5, when y = 3?

19   Write **one** multiple of 4 that is bigger than 135 but smaller than 145

20   How much is double 17.7?

	Test 8b Answer Sheet	
	Help	Answer
1		
2	14	
3		
4	100	
5		
6		
7		
8	4200	
9	80	
10	7.5	
11	25%	
12	7:50a.m	
13	1.5	
14	7°C	
15	£3.30	
16	45°	
17	22 & 72	
18	3y + 5	
19		
20	17.7	

	Test 8b Answer Sheet	
	Help	Answer
1		
2	14	
3		
4	100	
5		
6		
7		
8	4200	
9	80	
10	7.5	
11	25%	
12	7:50a.m	
13	1.5	
14	7°C	
15	£3.30	
16	45°	
17	22 & 72	
18	3y + 5	
19		
20	17.7	

Test 8c

1   How many threes are there in twenty one?

2   Find the total of 12 and 18

3   What is the value of 6 x 8?

4   Share two hundred between four

5   How many days are there in August?

6   What is the square root of one hundred and forty four?

7   Write down the prime number that is bigger than 20 but less than 25

8   How many sevens are there in two thousand eight hundred?

9   What is the answer when sixty is added to one thousand and fifty three?

10   What is 6.2 divided by 10?

11   What is 25% of £9?

12   What is 3 hours and 35 minutes after 6:50a.m?

13   What is 1.6 multiplied by 3?

14   A temperature of 7°C is recorded. The temperature drops by 20°C.
    What is the new temperature?

15   Twelve cakes cost three pounds and fifty pence.
    How much do six cakes cost?

16   One angle in a triangle is 55° and another is 85°
    What is the size of the third angle?

17   What number is exactly half way between thirty three and eighty three?

18   What is the value of 3y - 5, when y = 3 ?

19   Write **one** multiple of 4 that is bigger than 205 but smaller than 215

20   How much is double 15.9?

	Test 8c Answer Sheet				Test 8c Answer Sheet	
	Help	Answer			Help	Answer
1			1			
2	12		2		12	
3			3			
4	200		4		200	
5			5			
6			6			
7			7			
8	2800		8		2800	
9	60		9		60	
10	6.2		10		6.2	
11	25%		11		25%	
12	6:50a.m		12		6:50a.m	
13	1.6		13		1.6	
14	7°C		14		7°C	
15	£3.50		15		£3.50	
16	55°		16		55°	
17	33 & 83		17		33 & 83	
18	3y - 5		18		3y - 5	
19			19			
20	15.9		20		15.9	

Test 8 Answers

	Test 8a Answers	
	Help	**Answer**
1		8
2	13	40
3		56
4	100	25
5		30
6		5
7		17 or 19
8	3500	500
9	70	3116
10	4.9	0.49
11	25%	£1.25
12	7:55a.m	10:10a.m
13	1.4	4.2
14	6°C	-9°C
15	£3.20	£1.60
16	45°	70°
17	24 & 74	49
18	4y + 5	17
19		148 or 152
20	18.6	37.2

	Help	Answer
	Test 8b Answers	
1		9
2	14	50
3		72
4	100	20
5		30
6		9
7		11 or 13
8	4200	600
9	80	3121
10	7.5	0.75
11	25%	£1.50
12	7:50a.m	10:15a.m
13	1.5	4.5
14	7°C	-8°C
15	£3.30	£1.65
16	45°	50°
17	22 & 72	47
18	3y + 5	14
19		136, 140 or 144
20	17.7	35.4

	Test 8c Answers	
	Help	Answer
1		7
2	12	30
3		48
4	200	50
5		31
6		12
7		23
8	2800	400
9	60	1113
10	6.2	0.62
11	25%	£2.25
12	6:50a.m	10:25a.m
13	1.6	4.8
14	7°C	-13°C
15	£3.50	£1.75
16	55°	40°
17	33 & 83	58
18	3y - 5	4
19		208 or 212
20	15.9	31.8

Test 9

Test 9a

1   What is the sum of 3, 4 and 7?

2   What time is it half an hour before 11:15

3   How many nines are there in 72?

4   What temperature is ten degrees warmer than minus three degrees Celsius?

5   Take 13 away from 30

6   What is 325 divided by 25

7   What is 20% of 300?

8   A train travels at 60km an hour for an hour and a half.
    How far does the train travel?

9   What is the total of 3.5 and 2.15?

10   What is double 66?

11   How many seconds are there in 3 minutes?

12   I spend £34.50. How much change do I get from £100?

13   A bottle has 365ml in it. How many ml of water do I need to add to make it up to a volume of 1.5 litres?

14   What is three-quarters of 100?

15   What is double 6.9?

16   What is the mean of 4, 16 and 7?

17   An isosceles triangle has one angle of 35°. What could the other two angles? be?

18   The length of the base of a right angled triangle is 12cm and the height is 5cm. What is the area of the triangle?

19   If 2 pizzas are shared equally between 6 people, What fraction of pizza do they each get?

20   In a year 6 class there are twice as many boys as girls. There are 7 girls. How many children are there altogether?

	Test 9a Answer Sheet				Test 9a Answer Sheet	
	Help	Answer			Help	Answer
1			1			
2	11:15		2		11:15	
3	72		3		72	
4	10°C		4		10°C	
5	13		5		13	
6	325		6		325	
7	20%		7		20%	
8	60km/h	km	8		60km/h	km
9	3.5		9		3.5	
10	66		10		66	
11			11			
12	£100	£	12		£100	£
13	1.5 litres	ml	13		1.5 litres	ml
14			14			
15	6.9		15		6.9	
16			16			
17	35°	A = °	17		35°	A = °
		B = °				B = °
18		cm²	18			cm²
19			19			
20	7		20		7	

Test 9b

1 What is the sum of 4, 5 and 6?

2 What time is it half an hour before 10:15

3 How many nines are there in 81?

4 What temperature is ten degrees warmer than minus four degrees Celsius?

5 Take 14 away from 30

6 What is 225 divided by 25

7 What is 20% of 400?

8 A train travels at 70km an hour for an hour and a half. How far does the train travel?

9 What is the total of 3.5 and 1.45?

10 What is double 88?

11 How many seconds are there in 4 minutes?

12 I spend £29.50. How much change do I get from £100?

13 A bottle has 425ml in it. How many ml of water do I need to add to make it up to a volume of 1.5 litres?

14 What is two-fifths of 100?

15 What is double 7.6?

16 What is the mean of 6, 16 and 8?

17 An isosceles triangle has one angle of 40°. What could the other two angles? be?

18 The length of the base of a right angled triangle is 14cm and the height is 5cm. What is the area of the triangle?

19 If 3 pizzas are shared equally between 12 people, What fraction of pizza do they each get?

20 In a year 6 class there are twice as many boys as girls. There are 9 girls. How many children are there altogether?

	Test 9b Answer Sheet				Test 9b Answer Sheet	
	Help	Answer			Help	Answer
1			1			
2	10:15		2		10:15	
3	81		3		81	
4	10°C		4		10°C	
5	14		5		14	
6	225		6		225	
7	20%		7		20%	
8	70km/h	km	8		70km/h	km
9	3.5		9		3.5	
10	88		10		88	
11			11			
12	£100	£	12		£100	£
13	1.5 litres	ml	13		1.5 litres	ml
14			14			
15	7.6		15		7.6	
16			16			
17	40°	A = °	17		40°	A = °
		B = °				B = °
18		cm²	18			cm²
19			19			
20	9		20		9	

Test 9c

1 What is the sum of 2, 6 and 8?

2 What time is it half an hour before 09:15

3 How many nines are there in 108?

4 What temperature is ten degrees warmer than minus six degrees Celsius?

5 Take 16 away from 30

6 What is 275 divided by 25

7 What is 20% of 500?

8 A train travels at 80km an hour for an hour and a half. How far does the train travel?

9 What is the total of 4.5 and 2.55?

10 What is double 99?

11 How many seconds are there in 6 minutes?

12 I spend £27.50. How much change do I get from £100?

13 A bottle has 595ml in it. How many ml of water do I need to add to make it up to a volume of 1.5 litres?

14 What is four-fifths of 100?

15 What is double 8.6?

16 What is the mean of 4, 9 and 8?

17 An isosceles triangle has one angle of 50°. What could the other two angles? be?

18 The length of the base of a right angled triangle is 15cm and the height is 6cm. What is the area of the triangle?

19 If 3 pizzas are shared equally between 15 people, What fraction of pizza do they each get?

20 In a year 6 class there are twice as many boys as girls. There are 12 girls. How many children are there altogether?

	Test 9c Answer Sheet				Test 9c Answer Sheet	
	Help	Answer			Help	Answer
1			1			
2	09:15		2		09:15	
3	108		3		108	
4	10°C		4		10°C	
5	16		5		16	
6	275		6		275	
7	20%		7		20%	
8	80km/h	km	8		80km/h	km
9	4.5		9		4.5	
10	99		10		99	
11			11			
12	£100	£	12		£100	£
13	1.5 litres	ml	13		1.5 litres	ml
14			14			
15	8.6		15		8.6	
16			16			
17	50°	A = °	17		50°	A = °
		B = °				B = °
18		cm²	18			cm²
19			19			
20	12		20		12	

Test 9 Answers

	Help	Answer
	Test 9a Answers	
	Help	Answer
1		14
2	11:15	10:45
3	72	8
4	10°C	7°C
5	13	17
6	325	13
7	20%	60
8	60km/h	90km
9	3.5	5.65
10	66	132
11		180
12	£100	£65.50
13	1.5 litres	1135ml
14		75
15	6.9	13.8
16		9
17	35°	35° &110° or both 72.5°
18		30 cm²
19		1/3
20	7	21

	Test 9b Answers	
	Help	Answer
1		15
2	10:15	9:45
3	81	9
4	10°C	6°C
5	14	16
6	225	9
7	20%	80
8	70km/h	105km
9	3.5	4.95
10	88	176
11		240
12	£100	£70.50
13	1.5 litres	1075ml
14		40
15	7.6	15.2
16		10
17	40°	40° &100° or both 70°
18		35 cm²
19		1/4
20	9	27

	Help	Answer
		Test 9c Answers	
1		16
2	09:15	8:45
3	108	12
4	10°C	4°C
5	16	14
6	275	11
7	20%	100
8	80km/h	120km
9	4.5	7.05
10	99	198
11		360
12	£100	£72.50
13	1.5 litres	905ml
14		80
15	8.6	17.2
16		7
17	50°	50° &80° or both 65°
18		45 cm²
19		1/5
20	12	36

Test 10

Test 10a

1 What is a third of 24?

2 What is three thousand add twenty?

3 What is five squared?

4 How many millimetres are there in 1.5 litres?

5 What is 225 subtract 99?

6 What time is it 35 minutes before 2pm?

7 How much do you need to add to £3.65 to make £5?

8 What is 2 x 2 added to 3 x 3?

9 What is 203 added to 108?

10 What is 1.5 - 1.25?

11 A cube has sides of 3cm in length. What is the volume of the cube?

12 Pencil cases cost £2.30 each. How much do 3 pencil cases cost?

13 Two cakes are each cut into quarters. How many pieces are there?

14 Who many thirties are there in 600?

15 If n equals 5, what is 2n - 3?

16 What is 36 added to 67?

17 What is 33 add 34 add 35?

18 A rectangle has a perimeter of 24cm. One of the sides is 3 cm in length
What are the lengths of the other 3 sides?

19 Birthday cards cost £2.49 each. How many can be bought for £20?

20 Multiply 25 by 9

	Test 10a Answer Sheet				Test 10b Answer Sheet	
	Help	Answer			Help	Answer
1	24		1		24	
2			2			
3			3			
4		ml	4			ml
5	99		5		99	
6			6			
7	£5		7		£5	
8			8			
9	108		9		108	
10	1.5		10		1.5	
11		cm^3	11			cm^3
12	£2.30	£	12		£2.30	£
13			13			
14	600		14		600	
15	2n - 3		15		2n - 3	
16	36		16		36	
17			17			
18	24cm	Side 1: Side 2: Side 3:	18		24cm	Side 1: Side 2: Side 3:
19	£2.49		19		£2.49	
20	9		20		9	

95

Test 10b

1   What is a third of 36?

2   What is four thousand add ninety?

3   What is six squared?

4   How many millimetres are there in 1.3 litres?

5   What is 345 subtract 99?

6   What time is it 40 minutes before 1pm?

7   How much do you need to add to £3.25 to make £5?

8   What is 2 x 2 added to 4 x 4?

9   What is 213 added to 109?

10   What is 1.5 - 1.35?

11   A cube has sides of 2cm in length. What is the volume of the cube?

12   Pencil cases cost £2.25 each. How much do 3 pencil cases cost?

13   Two cakes are each cut into thirds. How many pieces are there?

14   Who many thirties are there in 360?

15   If n equals 5, what is 3n + 3?

16   What is 36 added to 77?

17   What is 22 add 23 add 24?

18   A rectangle has a perimeter of 24cm. One of the sides is 4 cm in length
     What are the lengths of the other 3 sides?

19   Birthday cards cost £2.24 each. How many can be bought for £15?

20   Multiply 25 by 7

	Test 10b Answer Sheet	
	Help	Answer
1	36	
2		
3		
4		ml
5	99	
6		
7	£5	
8		
9	109	
10	1.5	
11		cm³
12	£2.25	£
13		
14	360	
15	3n + 3	
16	36	
17		
18	24cm	Side 1: Side 2: Side 3:
19	£2.24	
20	7	

	Test 10b Answer Sheet	
	Help	Answer
1	36	
2		
3		
4		ml
5	99	
6		
7	£5	
8		
9	109	
10	1.5	
11		cm³
12	£2.25	£
13		
14	360	
15	3n + 3	
16	36	
17		
18	24cm	Side 1: Side 2: Side 3:
19	£2.24	
20	7	

Test 10c

1 What is a third of 27?

2 What is four thousand add thirty five?

3 What is seven squared?

4 How many millimetres are there in 1.2 litres?

5 What is 240 subtract 99?

6 What time is it 37 minutes before 1pm?

7 How much do you need to add to £3.95 to make £5?

8 What is 2 x 2 added to 5 x 5?

9 What is 233 added to 109?

10 What is 1.5 – 1.45?

11 A cube has sides of 4cm in length. What is the volume of the cube?

12 Pencil cases cost £2.40 each. How much do 3 pencil cases cost?

13 Two cakes are each cut into fifths. How many pieces are there?

14 Who many thirties are there in 390?

15 If n equals 5, what is 2n + 2?

16 What is 36 added to 88?

17 What is 23 add 24 add 25?

18 A rectangle has a perimeter of 24cm. One of the sides is 5 cm in length
What are the lengths of the other 3 sides?

19 Birthday cards cost £2.04 each. How many can be bought for £20?

20 Multiply 25 by 5

	Test 10c Answer Sheet	
	Help	Answer
1	27	
2		
3		
4		ml
5	99	
6		
7	£5	
8		
9	109	
10	1.5	
11		cm^3
12	£2.40	£
13		
14	390	
15	$2n + 2$	
16	36	
17		
18	24cm	Side 1: Side 2: Side 3:
19	£2.04	
20	5	

	Test 10c Answer Sheet	
	Help	Answer
1	27	
2		
3		
4		ml
5	99	
6		
7	£5	
8		
9	109	
10	1.5	
11		cm^3
12	£2.40	£
13		
14	390	
15	$2n + 2$	
16	36	
17		
18	24cm	Side 1: Side 2: Side 3:
19	£2.04	
20	5	

Test 10 Answers

	Test 10a Answers	
	Help	**Answer**
1	24	8
2		3020
3		25
4		1500 ml
5	99	126
6		1:25pm
7	£5	£1.35
8		13
9	108	311
10	1.5	0.25
11		27 cm^3
12	£2.30	£6.90
13		8
14	600	20
15	2n − 3	7
16	36	103
17		102
18	24cm	Side 1: 3cm Side 2: 9cm Side 3: 9cm In any order
19	£2.49	8
20	9	225

	Test 10b Answers	
	Help	Answer
1	36	12
2		4090
3		36
4		1300 ml
5	99	246
6		12:20pm
7	£5	£1.75
8		20
9	109	322
10	1.5	0.15
11		8 cm³
12	£2.25	£6.75
13		6
14	360	12
15	3n + 3	18
16	36	113
17		69
18	24cm	Side 1: 4cm Side 2: 8cm Side 3: 8cm In any order
19	£2.24	6
20	7	175

	Test 10c Answers	
	Help	Answer
1	27	9
2		4035
3		49
4		1200 ml
5	99	141
6		12:23pm
7	£5	£1.05
8		29
9	109	342
10	1.5	0.05
11		64 cm^3
12	£2.40	£7.20
13		10
14	390	13
15	2n + 2	12
16	36	124
17		72
18	24cm	Side 1: 5cm Side 2: 7cm Side 3: 7cm In any order
19	£2.04	9
20	5	125

Printed in Great Britain
by Amazon